AN ALTERNATIVE APPROACH
TO SPECIAL RELATIVITY

AN ALTERNATIVE APPROACH TO SPECIAL RELATIVITY

Thomas A. Orofino, Ph.D.

iUniverse, Inc.
Bloomington

An Alternative Approach To Special Relativity

iUniverse books may be ordered through booksellers or by contacting:

iUniverse
1663 Liberty Drive
Bloomington, IN 47403
www.iuniverse.com
1-800-Authors (1-800-288-4677)

ISBN: 978-1-4620-2350-9 (sc)
ISBN: 978-1-4620-2351-6 (e)

Printed in the United States of America

iUniverse rev. date: 06/02/2011

CONTENTS

Preface

This book was put together primarily to describe for the technical reader an alternative approach to certain aspects of the Theory of Special Relativity, an accepted theory attributed in its final form to the work of Albert Einstein in 1905. His work in part represented an extension and reformulation of the 17th century work of Isaac Newton on the equations of motion according to classical physics. The reformulation was needed in order to properly accommodate applications of Newton's equations in situations where the scientific observer of a physical process and the phenomenon he is observing are in motion with respect to one another. The relative motion addressed is not a part of the phenomenon, which may also involve other aspects of motion within it. The theory addresses the observation process itself. The motion between the observer and the phenomenon is essentially a complication in obtaining correct information about the physics being observed. Thus, in order to receive the information of the phenomenon, a signal, usually light, must convey it to the observer. Since information about the physical process being observed originates from different parts of the experimental locale, and at different times, it is necessary in one way or another to compensate for the time it takes each signal to arrive at the observer's position. In such instances, the legitimate information about the phenomenon and the elapsed time required for each signal to reach the observer would be intermingled.

Some typical applications of Special Relativity are in particle physics, where an observer may wish to study physical activity occurring within his apparatus and moving rapidly with respect to him.

The initial portions of the book include a description of the "Scientific Method," applicable to all science, and a summary explanation of Newton's laws of motion, both intended for the general reader. In the Relativity portions following, the general reader additionally is given a description of why and how relativistic situations differ from their classical counterparts, and thus why the guidance of Special Relativity is needed.

The later portions of the book are addressed to the scientific reader. They deal with the author's mathematical development of an alternative approach to kinematic aspects of Special Relativity, an approach that leads to exactly the same end-results as the treatments provided in the standard textbook presentations. However, it is hoped that the alternative provides some new insights in the understanding of relativity and, in particular, is free from the mysticism that has a tendency to arise in the traditional treatments.

I. THE OPERATIONS OF SCIENCE

1. How Science Works

2. Perception and Reality in Science

3. Science and Common Sense

1. How Science Works

In the early days of motion pictures many people thought actors more or less made up the plots and lines as they went along. Later, when we became more sophisticated, we could appreciate the systematic and cooperative roles of scriptwriters, producers and directors who upon completing a film make a permanent contribution, for as long as we wish to keep records.

The pursuit of scientific inquiry utilizes a highly developed intellectual and social system. Generally, the objectives are to add to the body of knowledge by building on the work of others. Often, the progress is through a parallel effort involving many contributors, not necessarily working together in a formal sense, but profiting from and adding to the progress in the course of meetings, publications, and individual communications within the scientific community.

The parallel system of science, functioning as apparently the human brain does, is far more productive than would be the case were each person working on his or her private projects. There are exceptions. From time to time lone scientists, breaking new ground, provide major new advances. To name a few examples, one could list Isaac Newton (laws of motion and gravitation), James Clerk Maxwell (laws of electricity and magnetism), Charles Darwin and Alfred Wallace (natural selection) and Albert Einstein.

The methods of science are not deductive, in the philosophical meaning of logic. That kind of reasoning is exemplified by the familiar paradigm "if A equals B and B equals C, then A equals C."

3

It produces no new knowledge, merely restating in another way what is already evident from the introductory information given. Deductive reasoning could be useful, for example, in a courtroom where a good lawyer might elicit a confession from someone, confronting him with a convincing accusation.

New knowledge is much more than someone's secret. It is best attained through inductive logic, the primary basis for progress in science. It is not always appreciated, however, that the inductive contribution is never an irrefutable fact – it is a premise. The following is an example[1]: "Every crow ever seen was black; therefore all crows are black." The uncertainty, which qualifies the deduction as a premise, is of course that, however unlikely, someday a crow that isn't black might be observed. Thus, in advancing a contribution in science an observation is made and generalized beyond the present proof available. The premise is always liable to refutation at a later date. Nothing in science is certain fact.

One of those refuted contributions was the Ptolemy theory of the earth-centered solar system replaced, about two thousand years later, by the Copernican model of the sun-centered system. Actually, we know now that's not strictly correct either, because the sun revolves about our (Milky Way) galactic center. However, the Copernican model is more useful for practical purposes.

The history of the solar system representations also illustrates an unavoidable deficiency of the scientific method, human subjectivity. In the times of Copernicus and Galileo there were clerical objections to any representation other than an earth-centered universe. One

4

great scientist, Johann Kepler (laws of planetary motion), strove mightily to respect the traditional religious views, while at the same time pursuing his scientific inquiry to its fruitful end. It is all well and good to apply inductive reasoning in attempts to advance knowledge, but sometimes politics, personal considerations, and just resistance to change temporarily distort and delay the process. Science, like the arts, and often in parallel with them, can be trendy too, with the fashionable style of the day preferred by some participants.

The scientific approach has worked quite well over the years. Basically, a new discovery, idea, or different way of looking at things is offered by a contributor to the entire world for scrutiny, a proposition to be reinforced, improved, or discarded, as time will tell. The contributions that survive join the ever-accumulating body of scientific knowledge. Unlike the case of movie-making, a contribution is not necessarily the end of the story, because it and the original idea as well are always in jeopardy.

Music, sculpture, the medical arts, and most endeavors rest on a foundation of basic procedures. They have their formal rules and their uniquely defined jargon, and they incorporate the collective wisdom of those who came before. Scientific work is no different. The process is termed the Scientific Method.

The Scientific Method may not be written down as a daily reminder, but is certainly acquired by scientists from the accepted practices in their fields. Basically, the method consists of the sequence: observe, confirm, communicate and generalize. Thus, a

5

biochemist, say, might observe in the course of routine work an unusual cell mutation.   He knows, from the available current knowledge in his field, that the mutation probably had not been observed previously. It may be of scientific interest, or at least is an enticing observation. Attempts to recreate the phenomenon would be appropriate. If the observation seems important, one also would wish to make it and any speculations about its origin and significance available to others, usually through publication. The exposure resulting might well stimulate the interest of other investigators to attempt to confirm the work in their own laboratories, and to offer some possible explanations for it. Usually, the original experimenter, and others, also would try to generalize the findings from the original observation. In our example, they might speculate that the observation is part of a larger phenomenon already known in their field, or that it suggests some new and useful applications in the field of cellular biology. Thus, new opportunities for research would be created.

The process of critique and honing of the new findings typically continues for some time. Perhaps, eventually, the new discovery is satisfactorily verified and accepted by the scientific community. Then it is available to become part of the educational curricula of schools and colleges.

However, with this and all other examples of advancements in scientific knowledge, scientists accept that the findings might not survive the test of time. No scientific result, no matter how convincing originally, is likely to escape modification or even being discarded, eventually. The original scientist who made the new discovery, and who may well have passed away long ago, would have accepted in advance that, although in his work an important discovery was made, in the long run it could be found to be incomplete or even wrong, and replaced by some better interpretation. A scientific contribution was made nevertheless.

2. Perceptions and Reality in Science

Lightning flashes followed in time by thunderclaps are events of eyewitness reality to the casual observer. Further investigation would establish the phenomenon as a single event, explainable by the difference in signal speeds of light and sound from a common storm center. A stick half-submerged at an angle in water seems to be bent just at the surface of the water. Of course, it is not really bent, only appearing to be because of an optical illusion.

Generally speaking, experimental science is conducted by making observations, usually through experiments devised by the scientist to systematically tease out the secrets of nature. The term "observation" is to be taken seriously, since the only way information may be acquired is through the physical senses of the experimenter, or instrumental extensions of them, followed by intellectual interpretation. Sometimes, it is apparent or suspected that the observation may only be an expression of a phenomenon of more fundamental significance. In those instances one does not take the observation literally, and a distinction is thus made between perception by the senses and an additional step toward identifying "reality." One may never know when reality, without the quotation marks, has been reached because a deeper significance is always a possibility.

Our experience with colors in visible light is another example in which the observation, scientifically, is misleading. White (visible) light is a mixture of colors, the colors of the rainbow, which in the

9

right weather conditions, is resolved into its components. When we observe an object to be the color red, for example, we know that it only appears red because the white light impinging on it has been partly absorbed by the object and only a portion of it, the red part, has been reflected back and perceived by our eye. The object actually does not possess color as an intrinsic quality – rather, its chemical constitution is such that when white light is shone upon it, and only then, are certain portions of the white light spectrum absorbed by it and certain portions reflected. In the absence of light, that is, in the dark, it not only exhibits no color but in fact possesses none.

Some observations related to buoyancy can be superficially misleading. The old saying about sinking like a lead balloon, for example, is not necessarily appropriate. Objects like wood in water and helium-filled balloons rising in the air "float" because the spaces they occupy in their surroundings, water and air, respectively, weigh more as surroundings than do the bodies using up that space. It is simply a question of one body, the object, displacing another body, the equivalent volume of surroundings, and the lighter one rising. So, in fact there are lead balloons that rise, and boats made of concrete that float. Lead balloons about two meters in diameter have been constructed and demonstrated[2]. Concrete boats were built in WWII as possible replacements for the conventional types. In both applications it was simply a matter of making the walls of the objects thin enough so that the weight of the surroundings that they displaced was greater than the weight of the objects filling that space. The principle of the lighter-than-air dirigible is the same. With these objects the interiors are filled with helium (or, as in the case of the ill-fated Hindenburg, with flammable hydrogen) gas. Since these light gases do of course still contribute to the weight, one might reason he would be better off simply by evacuating the interior of the dirigible. The problem is that the air outside the shell of the object exerts an (atmospheric) pressure that would then immediately crush the hull. The walls could be made stronger to withstand the pressure, but then the hull would be too heavy for buoyancy. The air carried around inside the fuselage of a cargo Boeing 747 weighs more than two tons, a necessary "dead weight." In all of the above examples, knowledge of the physics of the situation commands an additional step for a realistic explanation.

The stick in the opening example seems to be bent because of a well-understood physical concept of light transmission through media, in this case through water and then air, and is fully describable in terms of a property known as the refractive index of a medium. Thus, why the light rays bend on the way from the submerged stick to the eye is known down to some level of understanding. However, the nature of the light itself is not completely understood on a fundamental basis. Accordingly, we are left as always with an incomplete recognition of "reality."

3. Science and Common Sense

Science would be a lot easier to do and to understand if it were always "logical." Much of it is intuitive and the researcher is always reassured when his or her new discoveries seem to make sense. Difficulties arise when the results seem at odds with what might be expected, and one must then be especially careful in verifying the work.

Submarines are vessels designed for operation at great depths of the ocean, at which pressure from the surrounding water may be many times the normal value at the surface. The occupants of the submarine are inside what is known as the pressure hull. They experience normal atmospheric pressure conditions, safe from the large pressure on the outside intent on crushing the hull. This latter point, we might note in passing, is a characteristic of nature, which dislikes differences and tries to eliminate them.

To attain maximum depth capability with safety, submarine designers make the pressure hull in the form of a long cylinder with a cross-section as perfectly circular as can be constructed. The circular geometry is chosen because of a mechanical principle from the physics of statics. This states that stresses acting on the outside of a ring will express themselves through compression of its material in the circumferential direction, not by collapsing the ring. The technical term for the phenomenon is hoop stress. However, the slightest deviation from circularity invites catastrophic failure.

In the early days of this technology someone decided that perfect circularity could be verified in a completed pressure hull simply by taking inside of it a rod of length equal to the inside diameter of the cylinder, maneuvering the rod into all possible contacts, and ensuring that the ends of it just touched a given section on the opposite sides. This seems reasonable enough. Certainly a circle would pass such a test. It turns out, as one would express it in mathematics, the test is a necessary one for a circular cross-section to pass, but it is not a sufficient one to prove that the cross-section is a circle.

An entire class of geometric figures, known as Reuleaux triangles[3] after their discoverer, passes the rod test. Any one such figure, of which the circle is a limiting case, has the property that, say, if cut out of cardboard and held lightly between thumb and forefinger could be rotated into all possible positions without changing the finger spacing. That amounts to the rod test.

Reuleaux geometry is used in making rotating bits that can drill square holes and in construction of the Mazda automobile rotary engine. The simplest one of the figures looks like an equilateral triangle with bulging sides, and certainly would not be confused with a circle.

For a collection of physics applications yielding often-unexpected outcomes, the reader is referred to a book of such examples[4].

II. CLASSICAL PHYSICS

Physics is concerned with the working of physical phenomena throughout the entire universe. It typically involves mathematical operations and equations, which will be kept to a minimum in the present section.

Historically, among the most important statements of how the physics of material bodies works are the three Laws of Motion of the English scientist Isaac Newton (1642-1727). His findings are expressed in terms suitable for motion, and in terms of a material quantity physicists call mass. Mass, usually expressed in kilograms (Kg), is related to the everyday measure we call weight. In physics, mass is specifically defined as that quality of a material body that resists acceleration. The body, in resisting acceleration, is either stirred into motion from a resting state, or induced to acquire another velocity than the one it had before the acceleration. The acceleration in either case is the result of a force applied to the body.

Newton's First law states that a body, either at rest or in movement at some constant velocity, will maintain that condition forever, unless disturbed by an applied force. The intuitive understanding of force is essentially correct in the scientific sense. Like velocity, force can be either positive or negative, depending on the direction in which it is applied. When applied to a material body, force is positive if it increases the velocity of the body, that is, accelerates it, and negative, if it slows the body.

The notions of force and acceleration may be illustrated by the situation of a billiard ball rolling at what seems to be uniform velocity across a felt covered table. The ball is actually gradually slowing down, because of friction between the ball and the felt. The action of the friction represents the applied force, in this case, causing deceleration expressed with a negative sign. The friction would bring the ball to rest eventually, if the table were long enough. Another law of physics teaches that the energy represented by the rolling ball is conserved in its totality. Thus, if the ball is slowing down, losing energy, the energy lost has to be accounted for elsewhere. It appears as heat energy in the ball, from slipping slightly and creating friction. The ball is warming up a bit, and if some of that heat dissipates into the air, the temperature of the air also has to be included in the totality representing the original quantity of energy.

Newton's Second law states that when a force, given the symbol F, is applied to a mass m acceleration of the mass results, in quantity inversely proportional to the mass. In equation form this statement is written as $a = F/m$, where the symbol "a" represents the numerical value of the acceleration. The rearranged form of the equation, providing the same information, is $F = ma$, the usual expression of the second law.

Newton's Third law states that for every force there is an equal and opposite force. As with all of the laws, the third is a matter of everyday experience. Suppose one begins to move, that is, to accelerate a stationary supermarket cart. The cart begins to roll, less so for a given intensity of push if the cart has considerable mass. At the same time as the cart moves forward the person pushing it experiences a backward force, equal and opposite. The pusher would slide backwards were it not for the friction of his grip on the floor. Another example of the Third law is the recoil of a rifle shot. The bullet goes in one direction and the rifle in the other.

A physical quantity of special interest in the physics of motion is that termed momentum, given the symbol p. In its classical form of Newton's physics, momentum is expressed by the product of mass and velocity of the body exhibiting the momentum, the equation $p = mv$. Momentum plays an important role in development of the Special Theory of Relativity, which shows that its classical form must be modified. A mass in motion always possesses momentum. As noted above, it also always possesses energy E, indicated for example in the collision of a moving billiard ball striking a stationary one, and causing it to acquire motion.

Energy also is modified in the correct, relativistic interpretation of momentum, leading to the famous formula $E = mc^2$, where c is the speed of light.

III. SPECIAL RELATIVITY

1. Background

2. Introduction

3. Development

4. Relativistic Momentum

5. Relativistic Energy

6. A Geometric Derivation for Stellar Aberration

7. A Geometric Derivation for the Doppler Effect

8. Reality in Special Relativity

9. Conclusions

1. Background

Special Relativity (SR) has a long history, from Galileo in the seventeenth century, through the major advances of Einstein in the early twentieth century, to the present day, where certain applications and interpretations are still controversial. The basic concept treats the communication of information. The information is conveyed through light signals, from the source of the signal in one inertial frame, that is, a reference frame moving at constant velocity, to an observer in another inertial frame. It is the relative motion between the frames that requires introduction of SR techniques.

The presently accepted version of SR grew out of earlier notions that light (electromagnetic energy) was a simple wave, like sound or ripples in a pond, requiring a medium in order to propagate and thus, through its travel from source to observer, to communicate information. We know today however that light can propagate in a vacuum, as well as in certain other media.

The relativity concepts introduced at the beginnings of the twentieth century aroused a great deal of interest within the scientific community, and also among the general public, in their case, in part because of its spectacular but sometimes fanciful, interpretations. Today, SR is expressed in terms of the space-time model, with its abstract geometric representations, and with much practical utility.

The alternative approach to SR described here offers no new results in the applications of the theory. It contributes by offering a more intuitive model for light propagation, consistent with the accepted approach, which seems to meet all the necessary requirements.

2. Introduction

Our interest here is a rather different approach from both the customary space-time and the equivalent time-dilation method based directly on the Lorentz relationships. We derive a Proper-time expression for time-of-flight of a photon, from its emission in one inertial frame to its arrival at another, using the Lorentz transformation to specify the relationships between coordinates of the two frames. From the result, we demonstrate how our approach leads systematically to the familiar expressions for momentum, energy and other relativistic quantities.

According to the physics of Newton's day, time was universal, that is, the same time at any instant for all observers. Newton did know that the speed of light c was finite, and its approximate value, from the Roemer measurements[5] on periodic eclipses of the moons of Jupiter. For whatever reason, he chose not to consider that aspect further.

Today, a better interpretation of Newton's statement might be that it applies to proper time or, more specifically, to local time. Proper time, according to Einstein[6], is time recorded when the event, the observer and one of the synchronized clocks of his coordinate system are all situated at the same location. In such instance a physicist with an identical coordinate system, in a second inertial frame, whose clocks had been synchronized with those of the first, would accept this proper-time characterization. He himself,

25

however, would in general make a different observation, because as a result of his motion the event communication would reach him at a later time. In order to reconcile the two observations, the second observer would somehow have to apply a correction for the information lag implicit in his own result.

Information lag can of course occur even in static situations involving phenomenon and observer. Consider, for example, measurement of the length of a billiards table by an observer situated at one end. His method is to release at time zero on his clock a ball that travels at precisely known and uniform velocity u. He times with the same clock its visible contact at time t with the far end of the table, then calculates table length d as $d = ut$. His result is a slight over-estimation of the table length. In order to arrive at the exact value, he must correct for visual signal lag, thus producing the corrected value $d^* = ut/(1 + u/c)$.

In another version of the above example, suppose the billiards table had been bumped, and set into uniform motion v away from the stationary observer, just at the moment of ball release. The length calculation now becomes an exercise in Special Relativity. An additional source of error has arisen because the photon, in communicating the far-end contact, must now traverse both the table length and another increment of path length that has accumulated during its flight, due to the receding observer.

The photon in the above examples, one might state anthropomorphically, does not know upon its departure from the far end where or when it will be detected. Fundamentally, only the time recorded over the actual distance traveled is important in accurate determination of table length, for either the static or the SR situation. However, in order to find the correct result for the SR case, one needs an appropriate expression for time-of-flight of a photon between inertial frames. Neither the static formula for length above, nor the classical (Galilean) expression for frame separation at photon arrival $d/(1 - v/c)$, will after division by c qualify for this purpose. The correct path length value lies between the two approximations.

Generally speaking, the SR observer in the usual situations is interested in interpreting information originating from another inertial frame, as that information appears to him, Lorentz-modified (Appendix A). SR theory, time dilation or space-time, essentially accommodates the photon time-of-flight aspect through use of a measuring system, dictated by the Lorentz transformation, in which the observed value of c is always constant. If, as we maintain, the appropriate correction for time-of-flight, alone, distinguishes SR from the predictions of classical physics, then introducing an explicit adjustment for it in SR situations is a legitimate alternative to consider. We investigate that alternative here, in terms of Cartesian space and proper time. In this approach, constancy of c again is preserved, c being defined as actual distance traveled divided by elapsed proper time between the inertial frames.

27

3. Development

The Lorentz transformation is the essence of SR. It was re-derived by Einstein in the development of his two principles of relativity, equivalence of inertial frames and constancy of c in all observations. The transformation provides the relationships between all time and space coordinates of one inertial frame with all those of the other frame of interest. Accordingly, it can be utilized to derive the particular relationship between a time T_1, when a photon leaves the origin of inertial frame S, and a time t_2, when according to the remote observation of observer S, it arrives at the origin of frame S'. Both times are the observations of S, but only one, T_1, is in proper time. Capitalization of the time letter is used here as a distinction for cases where observations are so expressed. For any other times involved, designated here as lower-case time letters, the event has been indirectly observed. A primed quantity means as usual that it was measured by an observer in S', unprimed, by an observer in S, proper or otherwise in each case.

The result from application of the Lorentz transformation (Appendix B) for the observed time difference of arrival and departure of the photon from S to S' is

(1) $t_2 - T_1 = \beta T_1/(1 - \beta) = (L_0/c)/(1 - \beta)$

Where β is v/c and L_0 is the initial separation of frame origins when the photon is sent. A second result from the Lorentz analysis of the photon travel is the relationship between the S remote

29

observation t_2 and its (proper time) counterpart in S′, T_2'. This is given as

(2) $t_2 = \gamma T_2'$

where $\gamma = (1 - \beta^2)^{-1/2}$.

The use of Eq. (1) and Eq. (2) gives an expression for the photon time-of-flight, expressed exclusively in proper times. The result is

(3) $T_2' - T_1 = ([(1 + \beta)/(1 - \beta)]^{1/2} - 1)(L_0/v)$

Again, L_0 is the separation of the S/S′ origins at time T_1 recorded by observer S, when the photon is sent from his frame origin. Time T_2' is its arrival time at the S′ origin, as recorded by the observer there. Since both times are proper, the difference $T_2' - T_1$ expresses a travel time agreed to by both observers.

An equivalent expression, exactly equal to Eq. (3), is preferable for subsequent use. As it turns out, Eq. (3) is, more specifically, the proper time difference for the concurrent, identical change in the S/S′ origin separation during the photon transit. We may therefore multiply Eq. (3) by v, thereby expressing an additional accumulation of travel path length of the S′ origin relative to that of the S frame origin, add to that L_0, which in combination represents the total travel path of the photon, and finally divide by c to re-establish photon transit time on the right-hand side. This equivalent result is

(4) $T_2' - T_1 = (L_0/c)[(1 + \beta)/(1 - \beta)]^{1/2}$

It may be shown, by starting instead with the corresponding Lorentz expression (Appendix B, Eq. (B8)), that the right-hand side of Eq. (4) applies equally to the transit $T_2 - T_1'$, that is, to the case where the photon observed by S arrives from S'.

The form of Eq. (4) shows that the total number of waves distributed over the photon signal path is conserved for all β. Only the wavelength itself changes, filling up the modified signal path. It is as if the distance between source and observer at photon emission, likened to a relaxed spring, is stretched to accommodate the actual length traveled. An application of Eq. (4), with selection of the appropriate L_0 value, applies for any specified emission and arrival coordinates in the inertial frames, functioning as new origins, in either direction of photon flight.

The round-trip time follows from Eq. (4). The photon, having traversed L_0 and the extra path generated by the $[(1 + \beta)/(1 - \beta)]^{1/2}$ factor, travels upon reflection that distance again, plus an additional portion generated by a second application of the factor. The result for the S/S' round trip is thus

(5) $T_3 - T_1 = (2L_0/c)[(1 + \beta)/(1 - \beta)]$

We note the multiplying factor of L_0/c, in each of Eqs. (4) and (5), is the corresponding Doppler shift[7]. These factors generate the travel distance adjustments and thus the transit times of the photon signals.

The aforementioned distinction between proper times T and remotely observed, transformed times t is also the basis for a distinction between relativistic and Galilean interpretations of

31

physical phenomena. The observed time-of-flight from S to S' for a photon expressed in Eq. (1) is, according to observer S, simply that provided by the classical formulation. A classical observer assumes stationary status, attributes all motion to S' and simply allows for the accumulating path length of the photon according to the usual geometric process. As with Newton's physics, no distinction is made between t and T forms of time observed for an event. The relativistic counterpart of Eq. (1), Eq. (4), is the correct version, as revealed by the Lorentz transformation. Therefore, we may deduce a useful expression for conversion of photon travel time for a given event pair from the classical to the correct relativistic interpretation, in either direction, by the ratio of Eq. (4) to Eq. (1). The result is

(6) $(T_2' - T_1)/(t_2 - T_1) = (T_2 - T_1')/(T_2 - t_1) = 1/\gamma$

We note that all of the times in the above conversion ratios either are, or can be treated as, proper. Therefore, for example, emission times for both versions can be equated to zero, resulting in arrival times representing times-of-flight.

4. Relativistic Momentum

We begin with the customary synchronization of clock times as the coordinate origin of inertial frame S′ overtakes and coincides with that of inertial frame S. We additionally stipulate that the frames accommodate nominally identical mass elements m at their origins.

The X and X′ axes may be considered to overlap for any selected span, so synchronization could just as well be accomplished at any abscissa locations that are provided with identical clocks. Another part of the synchronization process is measurement, by both S and S′ observers, of their relative velocity v. This is calculated by each, using the sliding motion of the other's axis, relative to his own, as measure of distance covered. The velocity confirmations are purely classical, since there is no information lag in the data collection process. Both observers agree upon the value found for v.

After synchronization is complete, relativistic distinctions may now become apparent whenever one observer, say S, must resort to indirect means of acquiring information originating in S′. Observer S proceeds to measure the momentum of the mass element in S′. First, he determines a proper origin displacement increment ΔX from his own selected clock readings ΔT as

(7) $\Delta X = v\Delta T$

This displacement is proper, because observation location is where the distance change is occurring, as accumulating time. Observation

33

by S of the corresponding time increment ΔT cannot be obtained directly. However, S can convert his indirect observation Δt to a proper time increment by use of Eq. (6). The S observation of momentum, expressed consistently in proper, relativistic units[8], thus becomes

(8) $p = m(\Delta X/\Delta t) = m\gamma(\Delta X/\Delta T) = m\gamma v$

In this final expression for momentum, observed by S, both displacement and time increments are stated in terms of the locations of the momentum events. With equal validity, S' measures at the same location his own, zero momentum.

The introduction above of gamma as a multiplier of the classical form for p is, technically, an observational artifact of information delay. It has an equivalent and deeper meaning also. The observation reveals to S that the S' mass is moving at a velocity greater than his, and thus is energized relative to him. The factor gamma appears here under inertial conditions that represent the same final situation as in non-inertial accelerator experiments, where the momentum form and the energy associated with it are established by direct measurement[9]. In that situation the particle properties are functions of the increased velocity relative to the laboratory, acquired by acceleration. In both situations, the parameter gamma is the statement of the first principle of SR, equivalence of the two frames of reference.

5. Relativistic Energy

The expression of relativistic momentum is sufficient for derivation of the corresponding properties of total energy and rest energy. We develop these anew here for continuity.

The work needed to arrest the momentum of the particle equals its kinetic energy, integrated over the interval from 0 to v. Since the velocity component of observed momentum is the source of the factor gamma, we may combine the two, forming the observed relativistic velocity (γv). Accordingly, we write for the kinetic energy

$$(9) \quad K = \int F dX = m \int [d(\gamma v)/dT] dX = m \int (v d\gamma v)$$
$$= m \int (\gamma dv + v d\gamma) v = m \int (\gamma v + v^3 \gamma^3/c^2) dv$$

where we have used $d\gamma = (v\gamma^3/c^2) dv$. Therefore we can state

$$(10) \quad K = - mc^2/\gamma + (m/c^2) \int (v^2)(v\gamma^3) dv$$

The remaining integral of Eq. (10) is in the form

$$(11) \quad \int U(v) W'(v) dv = m[\gamma v^2 - \int 2\gamma v dv]_0^v$$

The latter expression of Eq. (11) is the result of integration by parts. The fully integrated expression for K is thus

$$(12) \quad K = - mc^2/\gamma + m(\gamma v^2 + 2c^2/\gamma) = \gamma mc^2|_0^v = mc^2(\gamma - 1)$$

Kinetic energy is the difference between total energy E, including that contribution from motion, and the velocity-independent rest energy. Inspection of the result for K shows that, since γ is the only velocity-dependent quantity, the total energy is expressed by

(13) $E = \gamma mc^2$

The remaining quantity comprising kinetic energy in Eq. (12) has no velocity dependence, and may be designated the rest energy

(14) $E_0 = mc^2$

6. A Geometric Derivation for Stellar Aberration

Astronomical aberration is the optical effect revealing a difference between the inclination of a selected star and the axis of a telescope focused on it. The phenomenon arises from the orbital motion of the earth, and thus of the telescope, during the time a photon transits from objective lens to eyepiece.

The situations for observation of a star are shown in Fig. 1. Frame S is that of the earth upon which an observer with his telescope of length L_0 is viewing at some angle to the horizon. Frame S' is that of the surrounding atmosphere beyond the objective lens, through which the star photons reach earth at an angle ω' (the S' observation). An analogy is suggested by the previous thought experiment for momentum.

In the classical approximation, S' sends a photon to S, who times its passage through the telescope with his own clock. Distinction between t and T times do not apply in the classical case and the Galilean model that observer S uses for the photon transit is that of Eq. (1). That is to say, the observer makes the usual allowance for travel distance accumulation with travel time of the photon. The diagram for the situation is shown in Fig. 1, in dark line and with telescope angle θ_C. In a time Δt required for passage of the photon through the telescope, a length $c\Delta t$, its base moves a distance $v\Delta t$.

37

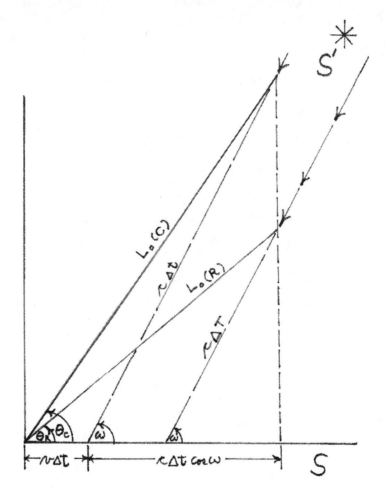

Fig.1. Classical and SR Aberration (See text.)

The angle θ_C is given therefore by

(15) $\tan \theta_C = c\Delta t \sin \omega' / (v\Delta t + c\Delta t \cos \omega') = \sin \omega' / (\beta + \cos \omega')$

38

The situation for relativistic aberration is that the photon requires a different transit time ΔT through the telescope, relative to distance moved along the base. Its passage is given by the lighter line in Fig. (1). Telescope length does not matter in these constructions. We find it convenient to choose a second, shorter telescope for the SR case, joining its objective lens end, the formal start of the photon transit, with the previous normal and using a shifted equivalent of the ray. We note that the displacement of the point where the normal intersects the X-axis is the same for both the classical and SR telescopes.

Since we are converting a classically-described time-of-flight to its relativistic form, we may invoke here the transit time ratio of Eq. (6). Only the ordinate portions of the telescope right-angle triangles differ, so we may combine the ratio with Eq. (15), thus specifying the relativistic angle θ_R as

$$(16) \qquad \tan \theta_R / \tan \theta_C = c\Delta T \sin \omega' / c\Delta t \sin \omega' = 1/\gamma;$$

$$\tan \theta_R = \sin \omega' / \gamma (\beta + \cos \omega')$$

We note that the two segments of the common abscissa of Fig. 1 each involve γ. The value of $\tan \theta_R$ in Eq. (16) for the case $\omega' = \pi/2$, that is, the overhead star case, is $1/\beta\gamma$, for all β. It contains the factor γ, even though the photon path and the path of the earth's motion are in this instance perpendicular.

The method used in the aberration derivation, adoption of Eq. (6) based on Eq. (4), implies equal treatment and results, whether relative motion is attributed to the star or to the earth.

7. A Geometric Derivation for the Doppler Effect

The Doppler shift observed by S, from a source of proper frequency N_0 situated at the origin of inertial frame S′, results from the light waves, at the moment of detection at the S origin, being distributed over a distance different from that at emission. Here, we specify that the paths of the two frames are parallel and, without loss of generalization, that all motion is attributed to S.

The array of emission paths generated by S′ is interrupted by the S observer and each path is characterized by a photon travel time which may be approximated classically or specified relativistically. Depending on the choice, different frequencies and angles of observation would result. It is as if S′ is provided with both a classical and a relativistic emitter.

For the origins separating and v positive as shown in Fig. 2, S will locate a particular optical path that corresponds to a classical interpretation. Just previously he would have found a shorter optical path, corresponding to the relativistic situation. For each situation the total number of wavelengths contained is conserved from emission to observation at the arrival inertial frame. The sum of the wavelengths in each case is the path length.

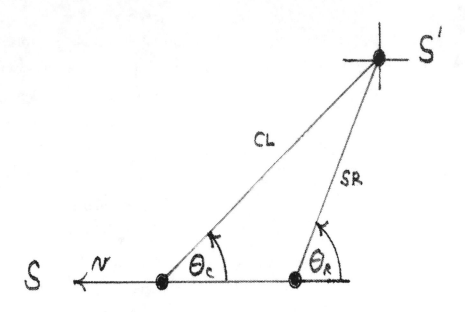

Fig.2. Classical and SR Doppler Effects (See text.)

For the classical approximation model, path length is given by Eq. (1) or $L_0/(1 - \beta \cos \theta)$ and therefore the frequency v_C observed by S would be

(17) $v_C = N_0(1 - \beta \cos \theta_C)$

This Galilean rendition may be arbitrarily assigned time-of-flight Δt.

In the case of the relativistic emission, the Eq. (4) path length is $L_0[(1 + \beta)/(1 - \beta)]^{1/2}$, L_0 having the same value as in the classical case, but a different angle θ_R and a frequency v_R observed. The time-of-flight corresponding to Δt is ΔT. The frequencies are inversely proportional to time and we may thus employ Eq. (6) to state that

(18) $v_R = \gamma v_C$

The combination of Eq. (18) and the modified Eq. (17) provides the expression for relativistic Doppler effect, namely

(19) $v_R/N_0 = \gamma(1 - \beta \cos \theta_R)$

At θ_R equal to zero, Eq. (19) reduces to

(20) $v_R/N_0 = [(1 \quad \beta)/(1 + \beta)]^{1/2}$

This last version represents the maximum effect, when X and X′ axes overlap.

8. Reality in Special Relativity

Is it real or is it just apparent? The reality of a remote observation, such as in our momentum derivation, is an arbitrary characterization. For example, a light signal, later Doppler-shifted as measured by the receiving observer, would be considered unambiguously real to the sender, but perhaps questionably so to the receiver, even though he measures a convincingly different energy of the beam.

Some definitions of reality are more informative than others. A suitable one might be the following. An observer stands at one spot, say in his laboratory, examining with his standard clock and meter stick a physical object or process. The object or process is then set forth on a journey, possibly subjecting it to forces, different environments and other circumstances which might change its nature. It is returned to the observer, at his same spot in the laboratory, and re-examined by him. The key criterion for establishing reality is whether the object or process changed in any measurable way. If it has, the conclusion is that real change occurred. Thus, for the Doppler shift above, the test would be to cancel the effect by reversing the motion of the receiving reference frame and then returning the light signal to the original sender. He would find it unchanged and conclude that the receiver's reality characterization was false, as far as our preferred observer, the sender, is concerned.

Two experiments in Relativity, which have been verified many times, are examples that directly address the question of reality and its consequences. First is the motion of a particle in the referenced accelerator experiment, in which the particle leaving the drift tube at a velocity very near the speed of light collides with a container equipped with a calorimeter. The perhaps physically damaged container and the record of the high temperature generated in the collision can be returned to the observer's laboratory spot and examined. The overall results show that all (real) consequences came from a non-classical identification of momentum, from which the energy is a result.

A second experiment[10], from General Relativity, is that of the effect of gravity fields on time rate. Identical clocks have been shown to accumulate and display the same elapsed times when placed side-by-side on the ground floor of a building. A difference in elapsed times was found, however, when one of the clocks was moved to a higher floor, left there for awhile, then in effect returned to the same spot next to the reference clock on the ground floor. The process qualifies as an example of reality according to the definition proposed above.

The two experiments cited lead to the conclusion (Einstein) that clocks exposed to gravitational forces, and clocks, say in S', which have been accelerated and are now moving at a velocity greater than that of the observer in S, are observed to run at slower rates than clocks left at the original laboratory spot. Of course, the "real" observation would be quantitatively different for another observer

46

moving at another relative velocity with respect to the clock of interest. In particular, to the observer in S', his clock rate has not changed at all. The situation is not unlike that of twins, separated at birth, one maturing in a foreign environment of unusual tranquility and standards of nutrition. He, being his own clock, might age more slowly than his home counterpart. Upon return, the traveler's conditions revert to the home circumstances, but the effects of advantageous maturation remain.

Momentum is thus real because we have no choice but to observe it with the mass element in motion with respect to the observer. The momentum results are purely observational, directly attributable to signal information delay, but are also an inherent part of the momentum concept, motion relative to the observer, induced by acceleration. The energized state of the photon may be considered a result of acceleration or of the velocity thus established relative to the observer.

In the case of time dilation evident under the influence of gravity fields, there is no relative motion between inertial frames. Presumably, the motion equivalent is that of the particles making up the physical clock being slowed by gravitational forces. We cannot, in this and all other, examples of time rate changes cited, attribute the observations to a change in "time itself." A tenet of physics is that one accepts only observations by the senses, or instrumental extensions of them. One must have a clock in one form or another.

9. Conclusions

Results from the alternative approach to SR described here provide some interesting insights. First of all, our basic premise is that the form of adjustment for photon travel time-of-flight in static situations must have a logical counterpart under dynamic conditions. In the static case discussed, agreement was provided between two observers, one observing remotely. If the observers are placed in uniform relative motion, nothing otherwise unusual has been introduced to the situation. Again, agreement between observers is to be expected, if the appropriate expression for information lag is used by the remote observer in order to correct his result.

The equation resulting for photon time-of-flight, expressed in times proper for the frames of the events, accepted by all, is developed from the Lorentz transformation, which relates all coordinates of one inertial frame with all coordinates of another. Accordingly, the result should be pertinent and reliable. In a sense, employment of our equation for the relativistic derivations is equivalent to conventional use of the time-dilation approach. In our case, the Lorentz operations are applied only once, the result then being applicable in a straightforward manner to all situations defined by light transits.

The equation for time-of-flight is quite reasonable in its form, namely original frame separation multiplied by the Doppler factor, divided by c. That form also appears for related, although quite different reasons, in sound propagation for the case when the

49

receiver is in motion and the source is at rest with respect to the medium. The conservation of number of wavelengths of the light signal shown by our equation is an interesting result.

Our equation expresses equanimity in favoring neither inertial frame, in keeping with the First Principle of SR. In an SR observation, a portion of the wave-like property of light is associated with the emitting frame.

The various origins of the factor gamma, as a distinguishing characteristic of SR, stem from a single phenomenon, accommodation of information delay in signal transmissions between equivalent, participating inertial frames. In the case of the momentum form γmv, indirect measurement is the evident explanation. In the case of the accelerated proton, the proximate explanation is the different energized state resulting in a different velocity than that of the observer. In both cases, the product mv could be measured directly, within the confines of the inertial frames involved, but that product is not momentum. Momentum and energy are inherently relativistic and necessarily involve two inertial frames. In their relativistic forms, the factor γ directly reflects the First Principle, by expressing a compromise for information delay accommodating both parties involved.

One could describe the passing of one inertial frame by another as a definition of velocity difference, thus implying that at some prior time of common rest frame status, one frame accelerated with respect to the other. According to Einstein, their observed clock rates now differ. Interpreted relativistically, the occurrence of an event in one frame, observed from the other, will reveal the difference and its consequences.

The foundations of SR are indeed demonstrated in the familiar station master/conductor thought problem[11]. To reprise, the two participants synchronize clocks at time zero as their positions overlap in the station, and further agree on a subsequent signal transmission. The stationmaster will send a light signal to the conductor in the train at a precisely designated time t_1 hence. The station master, attributing all of the relative velocity v to the train, expects the wave signal to arrive at the time $vt_1/(c - v) = \beta t_1(1 - \beta)$. The conductor, attributing all motion to the station master, expects to receive the wave in his train at time $t_1 + vt_1/c = t_1(1 + \beta)$. The actual arrival time is the geometric mean of the two interpretations in the thought problem, $[(1 + \beta)/(1 - \beta)]^{1/2}$, multiplied by L_0/c, that is, the expression of Eq. (4).

It would appear that the proper time-of-flight equation, although not in general describing actual observations, would suffice to correct all event data emanating from a distant inertial frame. Thus it would provide a basis for reproducing in the observer's frame all the details of a remote demonstration of universal physics. That accomplishment, in one way or another, is the goal of SR. The interpretation of gravitational forces according to Einstein's General Relativity theory also appears approachable through light signal delay considerations[12].

The classical, Galilean transformation is incomplete as an explanation of real-life observations. However, its predictions can be converted to the correct ones by the multiplicative factor $1/\gamma$, as described in our development. That interpretation of the role of γ, as a pedagogical issue, and in its practical applications, providing the correct relativistic forms for various quantities, are the principal results offered in this book.

IV. REFERENCES

1. The Random House Dictionary, Random House, New York (1966)

2. NBC TV (1977)

3. Sci. Amer. 208, 149 (1963)

4. The Chicken from Minsk, Y. B. Chernyak and R. M. Rose, Basic Books, New York (1995), ISBN 0-465-07127-9

5. Mathematical Principles of Natural Philosophy, Issac Newton, London (1687)

6. The Principle of Relativity, A. Einstein, Dover, New York (1952), p. 39

7. A. Gjurchinovski, Eur. J. Phys. 26, (4), 643 (2005)

8. Fundamentals of Physics, D. Halliday, R. Resnick and J. Walker, 5th Edition, Wiley, New York (1996), p. 974

9. W. Bertozzi, Am. J. Phys. 32, 551 (1964)

10. R. V. Pound and G. A. Rebka, Jr., Phys. Rev. Lett. 3, 439 (1959)

11. Encyclopedia Britannica, William Benton Pub., Chicago, 19 (1972), p. 94

12. The Feynman Lectures on Physics, R. P. Feynman, Addison-Wesley, Reading, MA (1963), Vol. 1, p. 7 – 11

V. ACKNOWLEDGEMENTS

The author wishes to acknowledge two members of the Department of Physics, Geology and Astronomy at the University of Tennessee at Chattanooga who were helpful to him in preparation of this book.

Professor Ling-Jun Wang introduced the author some time ago to the basic operations of the Lorentz Transformation, the centerpiece of Special Relativity. Thanks are extended for his patience.

The author particularly wishes to thank a long-time friend, Professor Eric T. Lane (retired), with whom he spent many weekly sessions discussing all aspects of physics, and eventually many other subjects as well. Professor Lane is particularly well informed in the basics of physics as it is accepted and taught in the standard curricula. Accordingly, he provided excellent guidance in keeping the book, if perhaps not free of speculative notions on Special Relativity, at least in reasonable consistency with the rest of physics.

Finally, the author wishes to acknowledge the proofing and editing contributions of Barbara Stagmaier in assembling the book.

VI. APPENDICES

Appendix A – Derivation of the Lorentz Transformation

Appendix B – Application of the Lorentz Transformation to Photon Time-of Flight

Appendix A – Derivation of the Lorentz Transformation

According to Einstein, the transformational relationships between coordinates (X, T) of inertial frame S and coordinates (X′, T′) of frame S′ must be linear, owing to homogeneity of space and time. Accordingly, the first space transformation may be expressed

(A1) $X = aX' + bvT'$

where v is relative velocity between the frames and a and b are constants. The proper time and space status of event coordinates in each case is emphasized by capitalization.

By application of the First Principle of SR, equivalence of inertial frames, we may write the second space transformation as

(A2) $X' = aX - bvT$

Elimination of X between Eq. (A1) and Eq. (A2) yields the first time transformation as

(A3) $T = aT' + [(a^2 - 1)/bv]X'$

Elimination of X′ between Eq. (A1) and Eq. (A2) provides the second time transformation as

(A4) $T' = aT - [(a^2 - 1)/bv]X$

A1

The speed of light c, measured independently within each frame and from each frame, must produce a common value that may be expressed in terms of the equations above as

(A5) $\Delta X/\Delta T = c$

$$= (a\Delta X' + bvT')/(a\Delta T' + [(a^2 - 1)/bv]\Delta X')$$

$$= (ac + bv)/(a + [(a^2 - 1)/bv]c)$$

(A6) $\Delta X'/\Delta T' = c$

$$= (a\Delta X - bv\Delta T)/(a\Delta T - [(a^2 - 1)/bv]\Delta X)$$

$$= (ac - bv)/(a - [(a^2 - 1)/bv]c)$$

Equating the right-hand sides of the Eqs. (A5) and (A6) yields

(A7) $a^2 = \beta^2 b^2 + 1$; $\beta = v/c$

We now may invoke the fact that the right hand side of Eq. (A1) is the optical path length observed by S, from his coordinate X to a coordinate X' in the S' frame, over which information is transmitted by photon propagation between these two points. The portions of the path within the frames and the portion between the frame origins, added together, constitute the total path length. As far as the photon is concerned, there is no distinction between the parts, and thus we may set a = b. The same argument applies to Eq. (A2). Thus, Eq. (A7) may be reduced to

A2

(A8) $a = b = (1 - \beta^2)^{-1/2} = \gamma$

The four Lorentz transformation expressions represented by Eqs. (A1), (A2), (A3), and (A4) with use of Eq. (A8) now become

(A9) $X = \gamma(X' + vT')$ $\qquad$ $X' = \gamma(X - vT)$

$\qquad$ $T = \gamma(T' + vX'/c^2)$ $\qquad$ $T' = \gamma(T - vX/c^2)$

Any two of the four equations above are independent. The other two can be derived from them.

Appendix B – Application of the Lorentz Transformation

to Photon Time-of-Flight

The Lorentz equations for S and S′ transformations, re-written from Appendix A, temporarily without Proper designations, are

(B1) $x = \gamma(x' + vt')$ $\qquad\qquad$ $x' = \gamma(x - vt)$

$\qquad$ $t = \gamma(t' + vx'/c^2)$ $\qquad$ $t' = \gamma(t - vx/c^2)$

We examine here the one-way transit of a photon sent from the S origin (Event I) to the S′ origin (Event II). In application of the transformations the observer always assumes stationary status, thus assigning all relative velocity to the other frame. The rules for wave transit apply and motion of the emitter of a light signal does not affect travel time of the photon. Assigned coordinates are $(x_1, t_1) = (0, t_1)$ and $(x_2', t_2') = (0, t_2')$.

For Event I, we have

(B2) $x_1' = \gamma x_1 - \gamma vt_1 = 0 - \gamma vt_1 = -\gamma vt_1$

and

(B3) $t_1' = \gamma t_1 - \gamma vx_1/c^2 = \gamma t_1 - 0 = \gamma t_1$

B1

For Event II, we have

(B4) $x_2 = \gamma x_2' + \gamma v t_2' = 0 + \gamma v t_2' = \gamma v t_2'$

and

(B5) $t_2 = \gamma t_2' + \gamma v x_2'/c^2 = \gamma t_2' + 0 = \gamma t_2'$

Event I and Event II yield by inspection the S′ observation

(B6) $t_2' - t_1' = (x_2' - x_1')/c = (0 - x_1')/c = \gamma v t_1/c$

Combination of Eqs. (B3), (B5) and (B6) yields

(B7) $t_2 - t_1 = \gamma t_2' - t_1$

$$= \gamma(\gamma v t_1/c + t_1') - t_1$$

$$= \gamma(\gamma v t_1/c + \gamma t_1) - t_1$$

$$= t_1(\gamma^2 v/c + \gamma^2 - 1)$$

$$= t_1(\gamma^2(1 + \beta) - 1)$$

$$= t_1(1/(1 - \beta) - 1)$$

$$= \beta t_1/(1 - \beta)$$

This last is written as $t_2 - T_1 = (L_0/c)/(1 - \beta)$ in Eq. 1 of the text, leading to the Eq. (4) expression for $T_2' - T_1$.

A similar analysis shows, for the S observation of a photon time-of-flight from S′ to S, that the equation corresponding to Eq. (B7) is

(B8) $t_2 - t_1 = \beta t_1 = \beta \gamma T_1$

The above equation in turn leads to an expression for $T_2 - T_1'$, which is equal to $T_2' - T1$ of Eq. (4).

ABOUT THE AUTHOR

I am a physical chemist by profession, Ph.D. Cornell University, 1956. My chemistry career interests were in the theory and experimental determinations of physical properties for high polymers. After retirement, I became very interested in physics, particularly in relativity theory. I invested a good deal of time in acquiring background in this new field, trying to understand it, in my own way. This book is the result. I have published one other book, in a safer subject area, titled "Sketches from Life and Travels."

Thomas A. Orofino

15 December 2010